I0468366

THE GOD QUESTION

Third Edition

by
John Edgell

A response to Richard Dawkins' *The God Delusion* and the Naturalist's hypothesis of existence without God

I have read apologists G.K. Chesterton, C.S. Lewis, J.P. Moreland, Francis Schaeffer, Ravi Zacharias, Lee Strobel, and others. Their arguments for the existence of God are cogent. For those wanting something more comprehensive than this work, I would recommend reading Schaeffer's *The God Who Is There*, and *He Is There and He Is Not Silent.*

This book is a rational look at the God question: Can we know that God exists? What evidence is there that God exists? This is written for the honest seeker of truth. Christ said, "Seek and you will find."

TABLE of CONTENTS

1. Contemplating the Possibility of God

*T*here are Theists, Deists, Pantheists, Atheists and Agnostics, and those who have no idea what they are. The Theist says that God is there, that he set the creation in motion, and is still involved in the creation. The Deist says that God is there, that he set the creation in motion and then, disinterested, removed himself from involvement in the creation. The Pantheist says everything is god. The Atheist (whom I will also refer to in this paper as a Naturalist) says that there is no God. The Agnostic (also generally a Naturalist) shrugs and says he does not know if there is a God. The basic question to begin with then is the question of the existence of God. Can we have confidence that God is really there?

In his book, *The God Delusion,* Richard Dawkins derides the God hypothesis (as he calls it). His position is that those who believe in God are deluded by superstition. That is about as meaningful as holding the position that if you like baseball you are a Yankees fan. Dawkins seems to think that unless you are a scientist, your ability to reason and the conclusions of your reasoning are of no value. In essence, to him science is *god* and scientists are the sole authority when it comes to discovering, knowing and expressing truth. Further, as far as he is concerned, theologians have no place in the discussion, and theology has no place in the educational system. He comes across as caustic in tone and uses ridicule to put that which does not agree with his *enlightened* viewpoint in its place. He speaks of other writers bullying their readers, but it seems to me he does the same thing, bullying any reader who might disagree with his position.

Dawkins attempts to refute the various proofs for the existence of God by mocking them rather than giving cogent rebuttals. Furthermore, the proofs that he attempts to refute are from past centuries— theologians such as Aquinas. He states that the only serious proof put forward today for the existence of God is the *seeming* appearance of design in nature. In reality, there are a number of other proofs, or reasons, to believe in the existence of God, other than the ones he attempts to refute with his caustic jibes. And in reality the proofs he does not address are of as great significance to the discussion as those he does address.

2. Earth's Meaningless Symphony

*I*n reading Dawkins, I must conclude that the wonder of all that exists is but a meaningless symphony. I find it interesting that one who is a materialist, believing that all things come from matter (he states this quite plainly in his book), would attack another viewpoint with such vehemence. If all things, including mind (and its thought processes) come from matter and are the result of some sort of natural selection, and if when we die we are simply "gone" and there is nothing more, then what I believe and do with my little piece of meaningless existence should be of no concern to him. According to his view, what I believe or do not believe does not make any difference to the eventual outcome, since the eventual outcome is non-existence for humanity and all life on planet earth.

Dawkins expresses the idea that by being good stewards we can perhaps prolong life on the planet. As an aside, I would pose the question: If we have a stewardship where does our stewardship originate? What does it mean to prolong life on our planet, and is prolonging life truly meaningful if it is all eventually going to come to naught? Reality: our sun will eventually expend its energy if we are not first swallowed by a Black Hole—so much for life on earth!

If it is all going to come to an end someday anyway, why worry about stewardship? Why not enjoy life now—no matter the cost to our planet—and let the processes move along as they will and be done with it? After all, some future generation must face the final score of earth's *meaningless* symphony of life. Selfish? Selfishness is a meaningless concept within a naturalistic construct, for nothing has any real meaning, all is absurd. Nature is the extent of things and there is no God, no root of accountability.

If there is no God, existence is absurd, telling me what I can and can't believe is absurd, social order is absurd. Death becomes its own Black Hole. Neither the individual nor society as a whole has meaning in the long view of the Naturalist. Meaning becomes an arbitrary thing without meaning. In essence, only what benefits me, or I perceive as benefiting me, would be of any value or meaning to me. And the only reason for me to treat you well would be if it in some way benefitted me.

3. The Naturalist's Hypothesis

Where does the naturalist's hypothesis take us as individuals and a society? Let us say I am impoverished and trapped in the inner city without the basics of life and without hope. In turn "society" tells me, "Here are the rules you must live by. Do not steal. Do not trespass on other's property. Do not harm others, etc." But why should I conform to those rules? The threat of society sending me to jail? Perhaps my daily life is worse than jail! If my philosophy (which the Naturalist has taught me) is survival of the fittest, I am going to do whatever it takes to survive even if it is not to *your* benefit or the benefit of society as a whole. The question is, "Is it to *my* benefit?" In other words, the Naturalist's viewpoint has no real value even if it were possible to prove it true. And surely, if there is a possibility that the Naturalist is wrong, then I would do well to explore that possibility, no matter how probable or improbable it may seem.

When it comes to figuring out whether something is probable or improbable, the process is open to manipulation. The outcome of that process depends on one's biases and the presuppositions upon which the probability factors are based. I have seen probability argued effectively in both directions.

On the other hand, if there is an alternative to the Naturalist's hypothesis, and that alternative makes sense of the whole, I would certainly choose the alternative. Why? Because the end product of the Naturalist viewpoint, which does not make sense of the whole, is absurdity—ultimate meaninglessness. Yes, in the moment we arbitrarily *attribute* some kind of meaning to life, but only because we cannot face non-meaning. And yet, one cannot get around the fact that non-meaning is the end of the Naturalist's atheistic hypothesis.

4. Examining the God Question: The Fact of Life

*N*ow, let us get to the God question. As I mentioned, there are numerous arguments for the existence of God. Some of the arguments carry more weight than others, but taken together one argument lends weight to the next. It is kind of like answering the question as to why a child should go to school. Some reasons will be more significant than others, but even the less significant reasons, while they might not have much impact on the question if they were to stand alone, still add to the weight of the more significant reasons.

One argument for the God hypothesis is the very existence of life. The Naturalist argues that *chance* conditions were *just right*, bringing certain chemicals and compounds together with the right level of heat, and life formed spontaneously. Sydney Fox's research on abiogenesis is used by some to support this assumption. It is claimed that he created "random and spontaneous" life-forms in the lab—proteinoids. While proteinoids are necessary to the simplest forms of life, they are not themselves life-forms. In that regard, as Dr. A.E. Wilder-Smith points out in his book, *The Genesis of Life*, the proteinoids created in the lab did not have any DNA coding, which the simplest life-form does have. Furthermore, all of Fox's attempts are at the expense of applied mental energy. Applying intelligence, he put together all of the intricate details in the lab, hoping to duplicate what *might have spontaneously occurred* in nature, where it is presumed that no intelligence was applied. And as noted, what he *created* was still a non-life form.

Fox's experiment is based on the fact that something exists, and there are two basic elements to that something, life and non-life. Either something exists or nothing exists. You cannot have it both ways. Again, by such experiments the Naturalists admit that something exists, and thus something has had to *always* exist. Since non-life does not become life, but life does become non-life, then the something that always exists must have life. Life as opposed to non-life argues for a Higher or Ultimate Life-form. So much so that some scientists, unable to explain how life evolved from non-life, espouse the theory that an asteroid struck the earth millions of years ago, and that asteroid impregnated the earth with a simple life-form that began the process of the evolution of the simple to the complex.

However, the asteroid answer is no answer. It accomplishes nothing to the aid of the Naturalist viewpoint. You still have the insurmountable problem of the origination of that impregnating life-form. Just coming from outer space does not make non-life to life more plausible. Still, on the one hand the scientist that espouses the asteroid theory should cause one to stop and give pause. Why? Because, in essence, what they are saying is that it is reasonable to believe that life had to be impregnated from outside nature as we know it, since Nature is simply material, so many atoms—non-life.

On the one hand, some Naturalists throw time at the problem. If you allow eons of time, the *magic* of evolution is possible. On the other hand, those who hold to the asteroid theory throw space at the problem.

Magic can happen somewhere out there in the vast reaches of space. But you cannot simply throw time or space at the problem of non-life becoming life and call it solved.

5. Examining the God Question: The Existence of Intelligence

Non-life to life argues for an Ultimate (outside) Life-form. Of course, God is more than just Ultimate Life. He is Ultimate Mind as well. Interestingly, the Bible says that man was made in the image of God—not only in the sense of life as opposed to non-life, but also in the sense of mind or intelligence—not only intelligence, but also the ability to reason—as opposed to mere sentient life. Both intelligence and reason are significant arguments for the existence of God.

We are wowed by the fact that a chimp can be taught to count and to choose one thing over another. But does a chimp say, "Humans are encroaching on my environment, I will rise up and save my kind!" No. It is only man who says, "Things are going poorly for this species and that species, as well as the planet. We must do something to save them." And indeed, we must! In Genesis one, God gave mankind a stewardship in that regard. But I do not see why the Naturalist should draw the conclusion that we have a stewardship toward any particular species or to save the planet. On what basis does the Naturalist have a stewardship? In fact, it seems to me that the idea of stewardship is contradictory to their thesis of natural adaptation and survival of the fittest. Would not a more consistent viewpoint be to let things happen as they will? The fittest will adapt and survive! If a certain beetle becomes extinct then it simply ran its course of existence and was not "fit" to continue to exist.

For example, in the winter of 2008, in the mountainous Ladakh region of Kashmir, 100,000 Pashmina goats faced endangerment and the possibility of death because of unexpected heavy snowfall that covered their pastures. In response, the government provided winter fodder. But what reason does the Naturalist have for saying, "How can we save this goat from possible extinction?" It is contrary to his hypothesis.

Consider the Himalayan, Tibetan antelope, the chiru. To protect this endangered species, shawls made from its wool have been banned. Survival of the fittest or survival of those we choose to help survive? Why do we act from outside nature to protect such species (and many others as well)?

Again, for the Naturalist, protecting such species is a contradiction, while for the Theist it is consistent with his or her theistic position. In Genesis 1:26-28 we read, "Then God said, 'Let Us make man in Our image, according to Our likeness; let them have dominion over the fish of the sea, over the birds of the air, and over the cattle, over all the earth and over every creeping thing that creeps on the earth.' So God created man in His own image; in the image of God He created him; male and female He created them. Then God blessed them, and God said to them, 'Be fruitful and multiply; fill the earth and subdue it; have dominion over the fish of the sea, over the birds of the air, and over every living thing that moves on the earth.'" Subdue and have dominion over does not mean to destroy and squander, but to rule over in a caretaking manner, to be stewards of the earth. Have Christians squandered? Yes. However, when they do so they are the ones being inconsistent.

The fact is, we humans interfere with nature's process destructively at times and constructively at times. When we see a given species struggling for existence, we apply intelligence and reason to the problem (whether it has anything to do with our own survival or not). We *reach in* (in essence, by putting ourselves outside of nature) and move things in a direction that hopefully will help save that species. You might say that we *play God*. And it is sometimes spoken of in those very terms. Why do we play God? Is it because we are the only species with the aptitude—the *image of God* capacity—to put ourselves outside and reach back inside? That should make us stop and ponder the God question.

Considering intelligence yet further: If intelligence, as opposed to mere sentient life, developed by means of the espoused principle of natural selection, then why is humanity the only species with pure intelligence? Am I really speaking of intelligence, or am I speaking of reason? I am speaking of intelligence—the capacity to acquire and apply knowledge. For instance, the savant has extraordinary intelligence in selected areas, and that intelligence is not tied to reason—the intellectual

28

faculty that correlatively processes knowledge and on that basis adopts actions to ends. Intelligence itself seems a contradiction to Naturalism, and the savant especially so.

Take the case of the Sudden Savant as set forth by the Wisconsin Medical Society on their website:

"A most amazing thing happened to me about 2 years ago at age 26 ½. While sitting in front of a piano at my friends' wine store at the local mall, just playing the keys, trying to figure out what's what and who's who, all of a sudden I felt things 'come together in my head'. It's a little hard to describe what went on in my head, but among other things, I suddenly (just like that) realized what the major scale was, what its chords were, and more importantly, where to put my fingers on the keyboard in order to play a certain part of the scale. The final ingredient in my gift, I noticed, was the ability to instantaneously recognize harmonies of the scales in songs I knew and reproduce them on the piano, as well as a somewhat less strong ability to reproduce melody by quickly figuring out how high or low the next note is, i.e. interval recognition."

"Suddenly, at age 26½, after what I can best describe as a 'just getting it' moment, it all seemed so simple. I started playing every song I knew from memory right there.

29

Suddenly people around and my friends stopped what they were doing, looked at me, and said 'whoa.....look at him play!!!' They, as well as I, were amazed at how just a few moments ago I was playing random chords on the keyboard, when all of a sudden I started playing like I had been a well-educated pianist."

Consider further the Polyglot Savant. Again, quoting an article found on the Wisconsin Medical Society website by Dr. Darold Treffert:

Most persons with savant syndrome have impoverished language skills as part of their basic disability, while musical, artistic, mathematical or mechanical skills flourish as particular islands of genius. Very rarely however, in an already rare condition, spectacular language (polyglot) skills, surprisingly, represent the island of genius in stark contrast to other overall handicaps.

Dr. Neil Smith of University College in London has been working with one such language savant—Christopher—for many years. In the book The Mind of a Savant: Language, Learning and Modularity, co-authored by Dr. Smith with Ianthi-Maria Tsimpli, Christopher's unique and prodigious language abilities—he can read, write and communicate in any of fifteen to twenty languages—are described in great detail.

Christopher was diagnosed with brain damage at age six weeks. Although walking and talking were somewhat delayed, at about age 3 an avid interest in factual books—telephone directories, dictionaries and books about flags or foreign currencies—developed, along with the ability to read not only in the usual fashion, but upside down or sideways as well. At age 6 or 7 Christopher showed interest in technical papers written in foreign languages that his sister brought home, fueling an obsession with languages that has lasted all his life.

In the 1995 book Christopher's language prowess is described thus: "He first came to attention because of his remarkable ability to translate from and communicate in any of a large number of languages. He has some knowledge (ranging from fluency to the bare elements) of: Danish, Dutch, Finnish, German, Modern Greek, Hindi, Italian, Norwegian, Polish, Portuguese, Russian, Spanish, Swedish, Turkish and Welsh." The book then gives examples of the breadth and depth of Christopher's expertise in those languages. Further, however, "like most professional linguists, Christopher can also identify languages from their written form without being able to speak or translate them, so he immediately, and correctly, identified Bengali, Chinese, Czech, Gujarati, Icelandic and so on, when presented with examples of them, and when given a postcard with 'thank you' written in a hundred languages on it, he identified twenty-nine of them."

Smith and Tsimpli go on to point out that the languages come from a wide range both genetically and typologically, and are written in a number of different scripts. Also remarkable is Christopher's ability to pick up languages quickly. They describe an incident shortly before Christopher was to appear on Dutch television: "It was suggested that he might spend a couple of days improving his rather rudimentary Dutch with the aid of a grammar and dictionary. He did so to such good effect that he was able to converse in Dutch—with facility if not total fluency—both before and during the programme."

How does he learn his various languages? Smith and Tsimpli indicate some have been gained as he "devoured" introductory 'Teach yourself' books. Other languages are picked up by interacting with native speakers and for others he has received explicit instructions.

I personally do not see how natural selection could possibly account for this kind of intelligence, and most certainly these elements of genius are not passed on today, individual to individual, by means of natural selection, even though it would be for the betterment of the next generation. Oh, one might say, it is not intelligence that is passed on by natural selection. How then did intelligence develop from non-intelligence?

Naturalism cannot effectively account for intelligence. Yes, intelligence rides the *fiber optics* of the brain. However, although the brain is matter (so many atoms), intelligence is not. And beyond the question of intelligence itself, is the human capacity to reason and to analyze.

As C.S. Lewis points out in his book, *Miracles*, "acts of reasoning are not interlocked with the total interlocking system of Nature as all other items are interlocked with one another. They are connected with it in a different way... The knowledge of a thing is not one of the thing's parts. In this sense something beyond Nature operates whenever we reason."

He goes on to say, "It is a matter of daily experience that rational thoughts induce and enable us to alter the course of Nature—of physical nature when we use mathematics to build bridges, or of psychological nature when we apply arguments to alter our own emotions." Then he notes that "rational thought is not part of the system of Nature. Within each man there must be an area (however small) of activity which is outside or independent of her. It does not follow that rational thought exists *absolutely* on its own. It might be independent of Nature by being dependent on something else."

Lewis also says that we must cry "Halt!" when the claim is made that reason comes from non-reason. He says, "If you don't, all thought is discredited." Then he adds, "It is therefore obvious that sooner or later you must admit a Reason which exists absolutely on its own." If any thought is valid, Ultimate Reason must exist as the source of our *less than ultimate* ability to reason.

For a more in-depth treatment of Reason as an argument for the existence of God, I would encourage you to read Lewis' book.

6. Examining the God Question: The Prevalence of Morals

Another argument for the existence of God is moral movement.

Consider: A coyote grabs your cat by the neck, kills it and steals it away, but in turn a fellow coyote sneaks up and steals the prey from the first coyote. Does either feel any remorse at killing the poor cat? Does either feel guilt at stealing from the cat's owner or from their fellow coyote? Do they sit down beside a tree in the wood and discuss whether their actions were right or wrong? No. Yet a number of years ago when I hit a deer with my car and had to kill the deer because its legs were broken, I felt remorse. If I were to steal from my neighbor I would feel guilt. And I have sat down with a friend and discussed right and wrong. These moral movements are not just the result of societal rules. Wolf packs have societal rules of sorts, but they do not have moral movements. The wolf pack does not deal in terms of right and wrong. Rather they deal in terms of Alpha male dominance and pack instinct.

On National Public Radio they were talking about the Kenyans protesting their recent election. One person said something about the crowds taking to the streets because they desired freedom. The leader of the discussion said, "But what about killing and slaughtering people of other tribes? What do you say to these same people burning a church with women and children inside?" Although they did not directly refer to moral right and wrong, the discussion was most certainly based on the mutually accepted judgment that burning a church with women and children in it was morally reprehensible—morally wrong. What made it morally wrong? Was it not the pack doing what the pack does? If wolves attacked another pack of wolves there would be no moral implication. It would just be wolves being wolves.

Only humanity deals in terms of good and evil, light and darkness, right and wrong. And atheists deal in those terms as well as those who believe God exists. The writings of Isaac Asimov are prime examples. His wonderful stories involve moral movement, good against evil, light against darkness. But if Naturalism is true, there is no basis for true moral movement. Good and evil, light and darkness, right and wrong are meaningless or at best given arbitrary meaning. Yet, all societies have the same core moral movements, the same or very similar basic concepts of right and wrong. They have core personal moral rights and wrongs and cultural or societal rights and wrongs. The societal rights and wrongs may vary widely, but again, the moral rights and wrongs are basic to all societies. From where does moral movement come? There is nothing in nature or in the Naturalist's hypothesis that gives a cogent explanation of moral movement. In fact, the radical Naturalist seems irritated by the very presence of moral movement, unless of course, it comes to someone stealing their property, threatening their life, or doing something contrary to "their" sense of right and wrong.

For instance, Dawkins decries all the evil done in the name of Christianity, while ignoring all the evil done in the name of atheism. It is a fact that more people have been killed, and/or oppressed, by atheistic regimes in the last sixty years than in the name of Christianity over the last several hundred years. But speaking of evil: Who decides what is evil? And if Dawkins or his ilk respond, "society," then no evil was done, because those in charge of determining what was right or wrong decided that what they did at the time was right. We look back and say, "No, it was wrong!" Without an Ultimate Law Giver, on what basis can the Naturalist say evil was done? And do not get me wrong, I believe evil was done, and it was evil because the Ultimate Law Giver says there is moral right and wrong.

It is interesting that on the Nature Channel we might watch a cheetah chase down a gazelle or a zebra, and although some may feel sorry for the poor gazelle or zebra, we do not say it is wrong for the cheetah to attack its prey. On the other hand, if a destitute man in New York mugs an old woman in order to have money for food, we say, "How awful!" Why? Do we say it is wrong because there is something higher in man, something noble, something more developed? But who determines man's response is higher, nobler or more developed? What makes man different? Natural selection? Moral movement does not come from natural selection. Moral movement is imposed from outside. Thus, when the destitute man mugs the old woman, he is condemned as a law breaker, whether he is ever caught or not. Moral movement speaks of a moral standard outside of man—an imposed moral standard. That is also the rub. We do not like things being imposed on us! We do not want God telling us what is right and wrong. Still, right and wrong stare us in the face every day!

Furthermore, the naturalistic hypothesis throws the door wide open to racism. It was the *evolutionary premise* that led Hitler to exalt the Arian race and justify the extermination of the Jews. I have had people tell me that whites are superior to other races using evolution to support their argument. Evolution leaves that door open. The Theist, on the other hand, believes that all people (of all races) are created in the image of God, and although not all people of any given race are equal in ability, they are ALL equal in value before their Creator.

7. Examining the God Question: The Genius of Creativity

*B*ut let us return once again to the existence of God. It seems to me that another proof for the existence of God is music and art, humanity's creative bent. Birds sing wonderful, but repetitious songs. You can distinguish one bird from another bird both by sight and by sound. Oh yes, some birds mock the sounds of other birds. But birds do not create new songs. And other birds do not sit around and clap their wings when they hear another bird sing. And why should man sing when he can talk? Why do we write poems when we can write essays? And why do we create symbolic art when we could copy the real thing? Why do we listen to another's song? Why do we read poetry? Why do we contemplate a work of art? From where does humankind's creative bent and the bent to appreciate another's creativity come? Certainly they do not come from mere nature. While singing so beautifully, birds have been known to sit and poop all over works of art. We take that in stride, but we would be outraged if men were to desecrate the same work of art. Why? Could it be that we see in art something beyond ourselves? And could that something be the creative image of God?

8. Examining the God Question: The Occurrence of Emotions

Another reality that puzzles me is laughter. A wolf may snarl, but you will never hear a good old belly laugh. Naturalism provides no basis for humor, for the wonder of a good laugh. The end product of naturalism is struggle, death and meaninglessness--nothing to laugh about. In meaninglessness there is no basis for laughter. Strange as it may sound, laughter speaks of the existence of God, because laughter speaks of meaning, of hope, of joy. On the other hand, the Naturalist Hypothesis leads only to a heavy heart and tears.

Can a Naturalist laugh and experience joy? Yes, but it is borrowed. The Naturalist's hypothesis certainly does not give any rational reason for joy—joy in absurdity? Joy in meaninglessness? Joy in no greater reality than death?

9. Examining the God Question: The Normality of Design

*N*ow to the question of design and Designer: Order as opposed to chaos. When I consider Dawkins' argument for lack of design it falls flat.

Is there chaos in the cosmos? Yes, but it is chaos in the midst of order, which is consistent with the Theist position. God created and His creation was good. In turn, chaos sullied the creation at the Fall—the fall of mankind in Genesis three, and the fall of Lucifer (Satan) as set forth in Isaiah 14:12-15 and Ezekiel 28:11-17. As a result of the Fall, the whole creation was thrown into chaos as noted in Romans 8:18-25

> For I consider that the sufferings of this present time are not worthy to be compared with the glory which shall be revealed in us. For the earnest expectation of the creation eagerly waits for the revealing of the sons of God. For the creation was subjected to futility, not willingly, but because of Him who subjected it in hope; because the creation

itself also will be delivered from the bondage of corruption into the glorious liberty of the children of God. For we know that the whole creation groans and labors with birth pangs together until now. Not only that, but we also who have the firstfruits of the Spirit, even we ourselves groan within ourselves, eagerly waiting for the adoption, the redemption of our body. For we were saved in this hope, but hope that is seen is not hope; for why does one still hope for what he sees? But if we hope for what we do not see, we eagerly wait for it with perseverance.

The astronomer, Kepler, was a Theist and believed God was the designer of the universe. It is told that he had a friend who was a materialist who believed it all *just happened*. According to the story, Kepler made a fairly detailed model of the solar system. When his friend came, he asked Kepler who had made the model for him. "Oh, it just kind of made itself," said Kepler. "Come now, who made it?" pressed his friend. Kepler responded, "You demand a maker of this poor model while you say that the original, that is far more complex and detailed, just happened without a maker?" Some question the veracity of the story, but whether it is fact or fiction it illustrates the point: Design demands a Designer.

In regard to design, the DNA coding system has no Naturalistic explanation. DNA code determines the *design* of each living organism. Codes do not just happen. There is absolutely no evidence for any kind of code coming about by accident. The very idea is irrational. Coding demands a Designer, a Code Encrypter, if you will. DNA demands God.

Evolution not only does not explain the presence of detailed encoding, it does not provide a realistic and non-contradictory means of moving from chaos to order, from non-life to life, from lower to higher. None of these phenomena are observable in nature. They are conjecture based on a biased presupposition that *this is the way it must have happened,* even though there is no evidence that it is happening that way today. We simply do not observe macro-evolution in the real world. We find such macro-evolution only in the *assumed* past.

The evolutionary concept goes against the grain of reality. The second law of thermodynamics tells us that without applied intelligence and applied energy, things move from order toward disorder, from higher to lower.

The fact is, for things to move from chaos toward order or from lower to higher, energy and intelligence must be applied. That kind of scientific reality *is* observable in the natural world. The Naturalist turns his or her back on law in order to create the myth of upward movement. If we give it enough time surely it could happen, even if we do not see it happening today, and even if it flies in the face of a known law of nature. And the very idea that there are laws of nature speaks for the Theistic Hypothesis of a Law Giver.

On the other hand, micro-evolution does take place and is consistent with the theistic and biblical thesis of God creating within kinds or individual species. There is development and adaptation to environment within those kinds, but there is not an evolution from one kind to another kind. And although such macro-evolution is presupposed by the Naturalist, we do not see any actual evidence of it in nature. So which hypothesis has the rational high ground?

10. Examining the God Question: The Pervasiveness of Worship

*T*he other day I heard someone comment on the fact (as they saw it) that there is an innate movement in man toward worship which extends across all cultures. And although there is a general rejection of dogma today, still, there is a broad emphasis on worship expression in terms of one's individual quest for some kind of spiritual experience. Even Carl Sagan (who did not believe in God) in his Cosmos series, stood in awe of the universe and declared worshipfully, "The Cosmos is god." It seems to me there is a god-like void in every soul and we all fill it with something, but does the content truly fit the void? And I must say that even the worship of God in terms of religion will not properly fill the void, because much of religion is god as we want him to be rather than God as he is. For many religious people their God is far too small, too boxed-in, too parochial, and too much of their own making.

Only the God of the Bible, when the Bible is interpreted as the straightforward plain message that it is, fills that god-like void satisfactorily. But such a God requires accountability, and accountability is what the Naturalist and many others make every effort to avoid.

11. Examining the God Question: The Validity of the Bible

*A*nd that brings us to the Bible, which is another proof of the existence of God. Christianity holds the Bible to be God's word, a message given by God to man. One might respond, "Yes, but there are many holy books that would make similar claims." Actually, most holy books claim to teach divine truth or contain divine wisdom, but they do not claim to be the very Word of God. "Oh, come on!" you say, "How can you make such a claim! A book is a book! The Bible was written by men!"

Yes, the Bible was written by men, men who in one way or another were given a message from God to mankind. The Bible was written over a period of 1,500 years by 44 different authors from tremendously diverse backgrounds, and yet the Bible is a consistent unit. When it speaks on history, geography or even science it is accurate, which stands unprecedented for the period in which it was written. The Bible speaks of life being in the blood (Leviticus 17:14; Deuteronomy 12:23), the earth hanging on nothing (Job 26:7), explains the water cycle (Job 14:11; 36:27-28; Psalm 135:7; Ecclesiastes 11:3), the wind currents (Job 37:11-12; 38:24), notes that the earth is round (Job 22:14; Psalm 8:27; Isaiah 40:22), and all of these long before they were scientifically understood. They were revealed truth rather than discovered truth.

Are there not contradictions and mistakes in the Bible? For years they said that the Bible was mistaken, that there was no historical record of the Hittites having existed. However, not that many years ago archeologists discovered that the Hittites were in fact a significant people. Dawkins points out that Quirinius was governor in Syria after the date when Christ was supposed to have been born, yet the Bible says that the census took place when he was governor in Syria. The fact is, he was governor at two different times and apparently was the de-facto governor at the time of Christ's birth (governing in behalf of the actual governor). Also, it appears that Augustus called for the census during Quirinius' first governance and that the census took place from region to region within the empire over a period of several years. The initial declaration was for each province to be registered, thus it was a local census that was part of a larger census. There is no contradiction, no error.

A SIDE NOTE: The manuscript evidence for the authenticity of the biblical text is overwhelming, and the latest manuscript evidence supports the gospels being written by the actual followers of Christ as indicated and at the time indicated. They in fact stand the test of being worthy, historical documents. To say they were written by other men and centuries later is simply outlandish, and is not based on the most up-to-date manuscript evidence, evidence that now firmly places the authorship of the gospels in the first century.

One can be confident that the Bible is God's Word, and one of the most convincing reasons for that confidence is fulfilled prophecy.

Manuscript evidence has now confirmed the book of Daniel was written before the fulfillment of the prophecies stated therein. Daniel foretold the Medo-Persian Empire, the Macedonian or Greek Kingdom and its ultimate division into four parts, and the emergence of the Roman Empire (Daniel 2:36-41; 8:8, 20-22). One prophecy that I found to be especially convincing was Ezekiel's prophecy concerning Tyre, a Mediterranean seacoast city. In Ezekiel 26 the prophet noted that Nebuchadnezzar (Babylonian Empire) would attack the city, and that other kings would as well, and that eventually another king would come along and not only destroy the city, but would scrape it into the sea so that where the city once stood would be a bare rock where fishermen would spread their nets. Many years later Alexander the Great fulfilled that prophecy in detail. To get to the island citadel some distance off shore he wiped the city into the sea, making a causeway by which to attack the citadel and destroy it. And as Ezekiel foretold it became a place for drying nets, and they still dry their fishing nets there today. And manuscript evidence validates that Ezekiel was written long before the time of Alexander the Great.

Although there are many such fulfilled prophecies I would mention one more that is extraordinary in detail and in its timing. It is found in Daniel 9:24-27.

"Seventy weeks **(as you study the text it is clear that the 70 weeks are weeks of years – 70X7 = 490 years)** *are determined for your people and for your holy city, to finish the transgression, to make an end of sins, to make reconciliation for iniquity, to bring in everlasting righteousness, to seal up vision and prophecy, and to anoint the Most Holy.*

"Know therefore and understand, that from the going forth of the command to restore and build Jerusalem **(this decree was given by Artaxerxes in 445 BC)** *until Messiah* **(Greek = Christos or Christ)** *the Prince, there shall be seven weeks and sixty-two weeks* **(62+7=69 weeks or 483 years)***; the street shall be built again, and the wall, even in troublesome times. And after the sixty-two weeks Messiah shall be cut off* **(Christ was crucified around 33 AD – It has been shown that the bishop who calculated the first calendar miss-figured by 6 years – 444+33+6=483)***, but not for Himself; and the people of the prince who is to come shall destroy the city and the sanctuary* **(The Roman General, Titus, destroyed Jerusalem and the temple in 70 AD)***. The end of it shall be with a flood, and till the end of the war desolations are determined* **(an**

undefined period of wars and desolations follow Messiah's death). *Then he shall confirm a covenant with many for one week; but in the middle of the week he shall bring an end to sacrifice and offering. And on the wing of abominations shall be one who makes desolate, even until the consummation, which is determined, is poured out on the desolate."*

Daniel foretold when Christ would appear on the scene and when he would be "cut off" (when He would die), and the book of Daniel was written hundreds of years before Christ, and the manuscript evidence is conclusive in that regard. For me, such prophecies confirm the Bible is indeed God's Word, and they are another proof of God's existence.

12. Examining the God Question: The Wonder of Love

*A*nother powerful proof of God's existence is the reality of love, of love and hate for that matter, for neither is organic to the natural world. A male and female dog mate, but they do not love in the sense that humans experience love. A wolf will kill a rabbit, but it does not hate the rabbit in the same sense that humans hate. Where do these experiential realities come from? Hate, it seems to me, is only explained by love fallen. Love is only explained by the existence of Love.

Love like reason comes from outside of man and is not connected to survival of the fittest and or natural selection. It is *other than*, and it transcends our *animalness* (if I may coin a new word). There is non-life. There is life. And there is image-of-God life. The Bible says that "God is love." We are told in John 3:16 that God so loved the world that He gave His Son to die for fallen mankind. Hate in turn is a result of the fall—love gone bad.

A mother bear will *instinctively* protect her young. That is natural to the natural world. But true self-sacrifice, which is the nature of love, such as a sister giving of her time and means that a sibling might go to college while she continues to work a dead-end job, is not at all natural to the natural world. In fact, that kind of sacrifice is contrary to how nature functions. Love screams to me that God is there!

13. Examining the God Question: The Impact of Experience

Another confirmation of God's existence is Christian experience.

One might ask, but is individual experience of God a viable proof of His existence? My response is, added to the other proofs, yes. Here again, Richard Dawkins mocks the idea as being ludicrous and uses examples that indeed are ludicrous and would be questioned by most thinking Theists. But Dawkins' straw men are the wrong straw men.

Experiencing God is not hearing voices or seeing visions, rather it is seeing, over time, the consistent faithfulness of God to the promises He has made in the Scriptures, experiencing the reality of His truth in one's life moment-by-moment, day-to-day. Promises such as, "Seek first the Kingdom of God and His righteousness and all these things (daily needs such as clothes, food, etc.) will be added unto you," and over the years God has faithfully met my needs, and at times in extraordinary ways. Furthermore, experiencing God is a life inexplicably changed by His power, love and presence. In that regard, for years I tried to change myself and it didn't take. However, when I gave my life to Christ He addressed the problem of my anger and accompanying temper and worked in my life to make me calm of spirit and gentle in demeanor. And whereas I was selfish, God has worked in my life to make me generous. Whereas I was deceitful, God has worked in my life to make me transparent and honest. That I have experienced the reality of God working in my life, for me is a powerful confirmation of God's existence, though it may only engender a questioning curiosity in the mind of the non-theist, or even dismissal. However, I cannot dismiss it. God's work in my life is reality.

God satisfies my hungering soul, while atheism leaves a meaningless void within, filled with contradiction and unanswered questions: Who am I? Where did I come from? Why am I here? Where am I going? Where did good come from? Where did evil come from? How can man be both noble and cruel? And on and on. In God I find honest, consistent, meaningful answers. Atheism is a *dead end* with no meaning, no hope, and no viable answers to the real questions of the soul. In fact, without God the whole game is absurd.

14. Examining the God Question: The Reality of Meaning

I find it interesting that the Naturalist believes we came from nothing, we are nothing (we have no meaning except whatever meaning we choose to give ourselves), and we will be nothing. The Theist believes that we came from Something, we are something (we have true meaning), and we will be something. To me the Naturalist position, that we came from nothing, is absurd. Nothing was there and somehow that nothing became something that at its essence is nothing, and although nothing became something, when we die we become nothing. Can you really speak of "nothing" of its own accord somehow becoming something in order to once again become nothing? Then nothing really exists. And yet I am here and so are you.

To ask the question, can God really exist, is no more absurd than to ask can my existence be real? That I exist, that anything exists, is mind-blowing, but the universe exists and I exist! That God exists makes sense of my existence.

The atheist/naturalist refuses rational evidence that God exists, but apparently they have no trouble believing there was nothing, and static nothingness "somehow" was magically energized by "something"— an extraordinary energy that did not exist since there was nothing. The energy exploded nothing into the vastness of the universe for no reason at all, and in the process again somehow magically rearranged the elements of non-life into life, and non-intelligence, into intelligence with the extraordinarily complex capacity to reproduce according to individual specie types.

God is plausible as well as possible, and gives meaning to all that is, but the notion that nothing becoming all that is, is irrational, neither plausible nor possible, and offers no meaning.

15. Examining the God Question: The Accountability Factor

*T*he real problem for most atheists is not the existence of God but the whole idea of being accountable to God if He does exist. Why else would they battle with such obsession against what they do not believe exists?

If God exists He is the ultimate authority and we are subject to him whether we want to be or not. We are ultimately accountable to him. However, fallen man does not want God telling them what is right and wrong and how to live their lives. That is the crux of the matter.

In essence man says, "I do not want to live as a rebel, with God and guilt hanging over my head. So I will do away with God and I can live as I please without guilt, without accountability to God." As a result they deny

God's existence, vehemently so! However, denial does not change reality. Because I deny something does not change reality. To deny the earth is round would not make it flat.

If we deny God's existence to avoid accountability to God it does not make us any less accountable. Yes, we can live as we please...but death takes us all. What then?

The Bible—God's Word to mankind—says in Hebrews 9:27-28: "And as it is appointed for men to die once, but after this the judgment, so Christ was offered once to bear the sins of many. To those who eagerly wait for Him He will appear a second time, apart from sin, for salvation."

Judgment or salvation provided by Christ when He paid the price of your sin and mine when he died on the cross. Which will it be? That is the ultimate question.

16. A Final Thought

*I*n the end there is either eternal Something or eternal nothing. Since something does exist there cannot be eternal nothing. Since nothing lacks meaning, and something does have meaning, then the Naturalist is wrong and the Theist is right. For if there is meaning it did not evolve from vacuous nothing. If there is meaning there is One who gives meaning. I accept Him by faith and yet my faith is not a leap in the dark. It is a step forward into light based on good evidence and sound logic. We long for unity within the diversity we see all around us. In the Trinity there is unity (One God), and diversity (Father, Son and Holy Spirit)...and they are in perfect community, a fellowship of love. Religion is not the answer. However, a living, vital, thoughtful faith in God, through the person of Christ, is the answer—loving relationship.

For the skeptic I would encourage the following prayer offered from a sincere heart: "God, if you exist, reveal yourself to me. Convince me."

Appendix

The Rub

A strange dichotomy tears at the fabric of philosophy in modern America—Britain as well. This philosophy does not throw aside the idea that "god" exists, and yet does not accept the existence of "God." The only god allowed existence in the corporeal world is a non-personal force within nature or the cosmos itself.

On the other hand, if one chooses to believe in a spiritual force of a strictly personal, non-scientific nature that appeals to the imagination, that "god" is acceptable. In other words, a "tribal god" is not a problem. However, your concept of god must remain on the opposite side of the divide from science. Once your god crosses the divide into true "God-ness" and claims some universality rather than merely being tribal…BANG! Your God must die! God is not allowed across the divide.

That is the reason gods of many kinds are endured throughout society. They are not a threat. They are mere products of the imagination. In fact, the scientific community has been known to flirt with the occult, as found in much science fiction. A good example is atheist Isaac Asimov's fiction. Yet, there is no such flirtation with the God of the Bible. Why? The reason is because the God of the Bible is bigger than science and encompasses all. Science will tolerate no competitor!

You see, the God of the Bible has no human limitations, not in intellect, capacity, or power. The God of the Bible is sovereign over man and over the entirety of what is—the universe and beyond. God is beyond comprehending, and that is the rub. The modern scientist/philosopher is unwilling to relinquish his authority to a God who unifies the whole of creation. Furthermore, if they accept the God of the Bible, the God who is bigger than the creation, it is incumbent on them to bow before Him as well as to acknowledge that He is far beyond their ability to comprehend."

With tribal gods man remains the central figure, the "knowing" one, and science reigns supreme. For that matter, science itself does not deny God. Men deny God, and they manipulate science to create an excuse for their denial. And for all who attempt to cross the divide…the sword falls! However, in the end God will have the last laugh. As noted in Psalm 2:1-4: "Why do the nations rage, and the people plot a vain thing? The kings of the earth set themselves, and the rulers take counsel together, against the LORD and against His Anointed, *saying,* 'Let us break Their bonds in pieces and cast away Their cords from us.' He who sits in the heavens shall laugh; the LORD shall hold them in derision." NKJV

God loves you and desires a personal relationship with you.

*T*he God Who is there, loves you and longs for relationship with you, for you were created in His image.

About the Author

John was born in Battle Creek, Michigan, and grew up in East Leroy, a small town south of Battle Creek. he graduated from Athens High School, served in the Army during the Viet Nam War era, and graduated from Appalachian Bible College and Berean Christian College, majoring in pastoral ministry, teaching and theology. He has been a Christian school teacher, a youth pastor, pastor, assistant mission director, mission director, and has been writing Christian fantasy fiction since his two sons were knee high to a gnome. He lives with his wife in Pahrump, NV.

See my website:

Johnedgell-author.com

Contact me at

jaedgell.author@gmail.com